The next step in
human evolution
has to be

the mastery of
what until now
has been

the unexplored
potentials
of the mind.

VALLEY OF THE SUN TAPE INSTRUCTION & IDEA MANUAL

Valley of the Sun Publishing, Box 38, Malibu, CA 90265

For A Current
Valley of the Sun
Tape Catalog
Of Hundreds Of
Life-Changing Programs
Write:

Valley of the Sun Publishing
Box 38
Malibu, California 90265

Published by Valley of the Sun Publishing Co., Box 38, Malibu, California 90265. Copyright 1981 and 1985 by Dick Sutphen. All rights reserved. No part of this book may be reproduced in any form without written permission from the publisher, except for brief passages included in a review in a newspaper or magazine. Printed in the United States of America.

ISBN Book Number: 911842-30-6

CONTENTS

Valley of the Sun tapes are produced in several different altered-state-of-consciousness formats:

- Deep-Level Meditation
- Hypnosis
- Meditation
- Hypnosis-Induced Meditation
- Sleep Programming
- Symbol Therapy

Each tape is structured to accomplish one of four things:

1

To assist you to overcome a specific habit, problem, fear or phobia. Examples: stop smoking, lose weight, eliminate depression, accelerate healing, let go of guilt or overcome jealousy.

2

To assist you to program and accomplish a desired goal, or to rapidly develop a new ability. Examples: to develop concentration, relaxation, speed reading, or to program ultra-monetary success or sports skill improvement.

3

To have mental experiences. Examples: Past-life regression, Higher-Self explorations, vision induction or astral projection.

4

To create a quiet level of consciousness in which you may receive meditative awareness.

1

VERY IMPORTANT

Do not use any altered-state-of-consciousness tape in a moving automobile for obvious reasons: The tapes are structured to put you into an alpha or theta level of sleep. They are very powerful and although you might feel that you could listen without effect, it's unlikely. Please use them in the appropriate circumstances.

Read the first four sections of this book before working with your tape. **This is absolutely essential!** You need this understanding to maximize the effectiveness of the tapes. "How To" instructions begin on page 34.

Before playing your tape, tap it firmly on both sides on a hard surface to release any tendency to internally bind. Most record stores do this automatically when a customer buys a tape.

The tapes use the most powerful techniques known to man and have been successfully used by thousands of people over the years. They are highly effective in assisting you to accomplish what you desire, if you are clear about exactly what you want. You can put the unlimited power of your subconscious mind to work for you **if you are willing to use the tape once a day every day until you accomplish the desired results.**

We will always replace any defective tapes.
However, when a tape player pulls the tape into
the machine and tangles it, it's usually because
the tape player is dirty or out of alignment. All
tape players must be thoroughly cleaned and
regularly demagnetized. Clean the heads,
capstan and rollers with a product that doesn't
destroy the rubber rollers. Your record store can
advise you.

It is a violation of the law to duplicate the
tapes for anyone else. In addition, it robs us of
income and the performer of royalties. So the
karma system applies to duplication—"As you
sow so shall you reap." On a subconscious level
the human mind is so ethical that it will always
balance every situation, regardless of how
inconsequential it may seem. In the "Idea"
section of this book we explain techniques in
which you would alter or combine tapes by
rerecording for your own personal use. This is a
different matter and is fully acceptable.

If you aren't achieving the desired results
with your tape read the **Problems** section
beginning on page 76. It covers the primary
difficulties that are occasionally encountered.

2

AN EXPLANATION OF
ALTERED STATES
OF CONSCIOUSNESS

13

The use of altered states of consciousness for positive programming, mental exploration or relaxation is perfectly safe and can benefit anyone. Altering your state of consciousness through hypnosis or meditation is the ultimate means of heightening your motivation to achieve your goals by directing your subconscious mind to work in cooperation with your conscious desires.

You are already familiar with altered states, although you may not realize it. You pass through these states at least twice a day, as you awaken in the morning and as you fall asleep at night.

Altered states of consciousness can be explained this way:

Brain researchers and medical practitioners have divided brain-wave activity into these four levels, based on cycles-per-second of activity:

Beta:	Full Consciousness
Alpha:	Falling asleep at night
	Awakening in the morning
	Hypnosis
	Meditation
Theta:	Early stages of sleep
	Deep hypnosis
	Deep meditation
Delta:	Full sleep to deepest sleep

Alpha, theta and delta are all considered altered states of consciousness. When you use hypnosis, meditation or sleep programming tapes, you will probably be in the mid-alpha range. In this stage you are definitely in an altered state, but you will also remain fully aware of all that is going on around you. If someone walks into the room while you are in an alpha-level altered state, you will hear them and sense their presence.

However, their presence will probably not affect you. In an altered state of consciousness, you tend to set aside the conscious mind (which still remains connected), and narrowly focus your attention on one thing. You can communicate directly with your subconscious, and suggestions given to your subconscious while in an altered state are nearly 100 times as effective in producing positive change as suggestions given in "normal" consciousness.

Important Facts About Hypnosis, Meditation and Altered States of Consciousness

1. Anyone who can concentrate for a few moments can learn self-hypnosis or meditation as a methodology for altering their state of consciousness. People with very low IQs, neurotics and very young children are not good subjects, for they usually have short attention spans. The best altered-state subjects **are** strong-willed, intelligent and imaginative.

2. There are many overlapping levels of altered states, but for the sake of simplicity they are broken down into three:

Light-Level Altered State: Your body becomes very relaxed, although you probably will feel you are not in an altered state. Most people achieve this level quite easily, and it is adequate for programming or explorations if you trust your own mind and are open to the impressions being received.

Medium-Level Altered State: You become relaxed to the point of losing awareness of your body. You are completely open to suggestion and are able to relive any suggested event. Although you remain aware, to some degree, of any outside disturbances, they will not distract you.

Deep-Level Altered State: You become almost entirely unconscious and will not be able to remember what transpires during the session unless specifically commanded to do so. One person in ten achieves this level; that person is called a somnambulist.

3. You cannot be controlled by the person who is guiding your altered state experience. Although many of you may have seen a stage hypnotist, for example, convince a man to do a beautiful, graceful hula dance, the hypnotist was not controlling the subject. Altered states of consciousness are also states of hypersuggestibility and typically, stage hypnotists choose to work with somnambulist subjects who are even more open to suggestion. Once hypnotized, the hypnotist simply expanded the subject's belief system for a moment. The subject chose to accept the suggestion and act upon it. Now, the man had seen a hula dance or observed one on TV or in a movie, so the knowledge of how to do it was locked in the memory banks of his subconscious mind. The hypnotist's suggestion expanded the subject's belief system that he could do the hula, thus he did.

4. Directly proposed suggestions cannot make you do anything against your morals, religion or self-preservation. If such a suggestion were given, you would either refuse to comply or would awaken.

5. The ego cannot be detached while in an altered state, so secrets will not come out, and you won't do anything you would not normally do if you felt comfortable about the situation.

6. You cannot become dependent upon altered states of consciousness because it has no permanent physical effect on the body or mind. However, it's true that many people look forward to their daily altered-state sessions because they become totally relaxed and awaken refreshed.

7. Self-hypnosis and meditation are learned techniques for achieving an altered state of consciousness. You do have to work at it, but if you are willing to use the first tape once a day for three weeks, you will become a conditioned subject. These tapes have been tested over several years, and incorporate the most effective induction techniques available anywhere.

8. As was shown in the diagram of brain-wave levels, both self-hypnosis and meditation can be used to achieve an altered state of consciousness. We offer both methodologies in our tape line.

3

TAPE TECHNIQUES

TAPE INDUCTIONS

Our hypnosis, sleep programming, deep-level meditation and meditation tapes offer several kinds of altered-state inductions.

Hypnosis Tapes: The hypnosis tapes offer four different inductions:

- A formal induction that uses counting techniques to suggest you go "down, down, deeper, deeper"
- A formal induction that uses counting techniques to suggest you go "up, up, higher, higher"
- An informal induction that uses various techniques such as visualizing a bright blue ball of light or walking through a forest
- A formal induction conducted by two people that uses counting techniques

Meditation Tapes: Gentle guided imagery is used to induce an altered state. This is often accompanied by nature sound effects or music You then will hear only the background sounds, allowing you to receive your own input.

Hypnosis-Induced Meditation: Relaxation and counting techniques are used to induce an altered state. You then will hear only the background sounds, allowing you to receive your own input.

Sleep Programming: The tapes begin with a body relaxation to assist you in falling into a normal sleep while you listen to the suggestions.

Deep-Level Meditation Tapes: A body relaxation technique is used in addition to instructing you to press your fingers to your psychic center.

TAPES TO OVERCOME PROBLEMS OR PROGRAM GOALS

Programming Techniques

Each of our programming tapes contain one or more of the following techniques to help you achieve maximum results.

Repeat Drive Technique: This technique condenses the primary concept of the tape into a powerful suggestion designed for maximum access of your subconscious mind, then repeats the suggestion over and over. You are asked to participate by mentally verbalizing the suggestion. The goal is to saturate the subconscious with a positive message about creating a desired change or achieving a desired goal. It is very effective in overriding old negative programming.

Mental Movies/Positive Affirmations: Mental movie is another name for cybernetic affirmations or the technique of visualizing your goal as an already accomplished fact. This is probably the most important aspect of altered-state-of-consciousness programming. Be sure to read this entire book to fully understand how and why the technique is so effective in helping to create positive change.

A mental movie is a fantasy of your own creation. The fantasy is a programming technique for the subconscious mind, which can't tell the difference between imagination and reality. You are the movie producer, director and actor. You use other people in your movies, but as the director you control their words, actions and reactions. You film this movie exactly the way you would like to create your own reality.

The key to successful programming is to make the mental movie as real as possible. See it with your inner eyes. Feel and sense everything that is happening—your emotions and the reactions of others who are observing your victory or accomplishments. Note facial expressions and other small details. Create the environment just as you would if you were actually producing a movie.

The movie always shows you achieving your objectives...viewing them as an already accomplished fact. This is important. By seeing your goals as already accomplished, your subconscious is able to fully comprehend the goal and immediately begin to work towards bringing it into reality. Because you are controlling communication while in alpha or theta brain-wave levels, you have maximized programming power.

Weight-Loss Mental Movie Examples

What follows are three examples of weight-loss mental movies:

1. You see yourself stepping on the scale after your morning shower. You watch the dial swing, slow down...and then stop at 115 pounds. You hear yourself exclaim with delight, see the smile on your face, feel the elation as you toss on your robe and run down the hall to tell your husband of your accomplishment. Next, create his reactions, etc.

2. You are downtown shopping and you run into a friend that you haven't seen for a long time. Your friend is amazed at your thin body and proceeds to tell you so. Create all the dialogue, etc.

3. You are in a women's clothing store and the saleslady asks what size dress you wear. "Size 8," you reply. Then you go through the process of trying the dress on and realizing it's a perfect fit. The saleslady compliments you on your figure, etc.

Create many different mental movies and run one today and another tomorrow, always varying and creating new ones to fit your situation and goal. Remember, you are limited only by your belief system.

Formulating suggestions for "think-in" programming: Some of the tapes leave open time in the middle for you to think-in your own suggestions. If you develop your altered-state abilities without the tape, you also need to be aware of how to formulate suggestions. First, you should write out your suggestions and become familiar with them. Then combine the technique of mentally talking to yourself (silently speaking the suggestion to your subconscious) and creating mental movies as I've already discussed.

Action can be easily inhibited by systematic thinking. Here's a simple example to illustrate this point. Pick up a pencil and hold it between your thumb and index finger. Now, begin to think "I am going to drop it. I am going to drop it." Now while you are **fully concentrating** upon the phrase, "I'm going to drop it," try to drop the pencil. It cannot be done. No matter how hard you try, you will be unable to let go of the pencil while you are focusing your mental energy on "I'm going to drop it," because you are concentrating on the future. Not until you send the message, "drop it!" from your brain to your fingers will you be able to let go of the pencil.

Thinking you can do something does not necessarily mean you can do it. This is how many people work at their goals, problems and hang-ups. The result is non-achievement and continued problems and hang-ups.

You must define your goal clearly, then set it into your subconscious mind as a positive, accomplished fact. Concentrate upon it and release it. Every day, concentrate upon it and release it. Don't worry about it, over-analyze it or doubt that you will achieve it.

24

You can change your life and create your own reality if you set realistic goals and then proceed towards them in a positive way.

Decide upon your goal and write it out clearly. The subconscious mind will work towards helping you accomplish your goal, but it has little reasoning power and seems quite dense when it comes to understanding. Write your goal many different ways and constantly repeat it. Memorize it so you can use your hypnosis time most effectively. Spend part of that time seeing your goal as an already accomplished fact. Do this in vivid detail.

As an example: You desire to achieve monetary security. Your suggestion might go something like this: "I absolutely have the power and ability to become monetarily secure. I can make money. Great sums of money. I create my own reality and from this moment on I begin to see opportunities to make money. Making money is a good thing and I am a money maker. I AM A MONEY MAKER. Monetary security is my reality." (Repeat and repeat.) Then visualize yourself with a savings account book in your hand. You are looking at the large figures posted in the book. It is your savings account book, your money. See your name on the front of the book, see your own happy expression...see everything about the situation in vivid detail.

Your subconscious mind will understand. It will begin to work for you. If you are willing to work at it, you can literally program yourself to any degree you desire. Everyone has heard of

25

the dog trained to salivate at the sound of a bell. You have the abilitiy to program yourself to this extent. Obviously, that is not your goal, but it makes the point.

Another extremely important thing to remember in formulating suggestions is: they must be positive. Often what we think of as positive is actually negative. If, for example, you go into an altered state and say, "I want to get rid of this headache...the headache is going away," you are actually programming yourself negatively and believe me, the headache won't go away. The word "headache" itself is a negative. Once again, the proper way is to concentrate upon the desired result: "I will awaken with my head feeling clear. My head will feel very good, relaxed and refreshed, when I awaken." (Note: Altered states should only be used to alleviate pain when you understand the cause of the pain. Pain is nature's way of telling you something is wrong. A headache is one thing, but a continuing stomachache could be appendicitis, etc.)

Conditioned-Response Trigger Words: On many of the tapes, trigger words are used for post-hypnotic response when you need an extra "hit:" usually concentration, motivation or energy. This is a standard programming technique that is amazingly effective once you have become conditioned and begun to use it. It's a good idea to wait until you've worked with the tape for three weeks before you call upon your trigger word programming. Then use it often, for each time you use a conditioned-response trigger word its effectiveness increases.

Example: The trigger phrase in the **Runner's Hypnosis Album** is "passing gear." When the hypnosis-conditioned runner is racing and feels the need for additional mental or physical stamina or a surge of energy, he will say these words to himself in his mind. The conditioned response programmed by the hypnosis tape then causes the subconscious mind to intensify the positive powers of the conscious mind and physical body.

To help yourself accelerate programming, visualize the trigger word as being totally effective when the suggestions are being given during the session.

TAPES TO HAVE MENTAL EXPERIENCES

HOW TO RECEIVE VIVID IMPRESSIONS

About 85 percent of those using self-hypnosis and meditation techniques to explore the unknown will easily receive vivid impressions. The remaining 15 percent must be convinced to trust themselves. They expect subjective input to be perceived in a particular way, and when their experience doesn't live up to the expectation, they block themselves. Their attitude is, **"I won't accept it unless it happens according to my expectations!"** Their belief literally destroys their experience, for they expect to receive perfect dreamlike impressions, and that simply isn't the way it works for most people.

There is absolutely no excuse for anyone not to receive. It is a simple process of self-trust. The experience is different for everyone, though, and this seems to be what some people find frustrating. The primary misconception is that the experience isn't real unless you can see pictures in your mind.

Though many people receive visual or fantasy-like impressions, others simply perceive thoughts or feelings. Some claim to see nothing at all and yet they are able to relate numerous details. Others get the impression that during the regression they are making it all up, even though later their experiences have been historically documented. You have to be willing to trust your mind and your impressions.

Most people perceive as if they were creating a fantasy in their own mind. Think about the last time you mentally relived an argument or experienced a sexual fantasy. You imagined the situation and became emotionally involved in it, yet you also remained fully aware of your surroundings; you realized you were creating the fantasy. Now stop and reread those last few sentences. They sum up the way you will probably feel while receiving in an altered state.

There are, of course, many other ways to perceive. For example, some people receive single pictures that shift like watching slides through a slide projector. Others hear a voice or read words. And each of you may perceive through any of these forms at any time. Sometimes you may be emotionally involved, as if you were actually reliving the experience, while at other times you'll be detached, perceiving the events as an observer.

Dick Sutphen has had hundreds of altered-state-of-consciousness experiences in which he has perceived vivid impressions, mind projections, past-life regressions and psychic input that has later been documented. Dick always believes at the time that he is making up the whole experience. "I receive a thought or fantasy impression just exactly the way most people do. The only difference is that I've learned to trust the input I receive in an altered state."

Dick is a light-level subject. In an altered state, he is fully aware of all that is going on around him. Yet he allows himself to trust the impressions. If you can trust yourself, you can't help but receive.

Let's carry the concept of impressions a little further. Dick claims that we all have a resident windbag in our mind who keeps up a running commentary on our every experience. He calls him "Babbler" because of his incessant, and usually meaningless, chatter. Take a few moments right now to get in touch with your own Babbler. Close your eyes and be quiet for about 30 seconds. Listen to the commentary going on in your mind. Do this now before you continue reading.

All right. Did you hear him babble? If you

didn't, Babbler was that little voice that was saying, "Voice? What voice? I don't hear any voice. This is ridiculous; there isn't any voice in my head!"

Unless you have experienced extensive Zen training to quiet your mind, or developed meditation to a great degree, Babbler probably never shuts up. Thus when you enter a state of meditation, the very least you can expect to receive is the sound of these babblings. It is impossible not to receive anything. The least you will receive is Babbler's babblings, and in an altered state they could be quite meaningful. Perhaps you'll be receiving valid data about your past or psychic data about your current or future reality.

Stanford Research Institute made the national news a few years ago when they proved that everyone is psychic if they will simply trust the impressions they receive. Under the auspices of the Institute, scientists Russell Targ and Harold Puthoff conducted extensive remote viewing (mind projection) experiments under laboratory-controlled conditions. The format of the experiment was to ask test subjects to psychically perceive distant locations that had been chosen at random. Test subjects were told: "You have permission to be psychic. In fact, we expect it of you." Test subjects were also told to trust the impressions they received, even though they would probably feel they were making it up.

The successful results of the study proved

conclusively that we have the ability to project our minds to, and accurately perceive, distant locations. It is not difficult to develop psychic ability. It can be difficult to trust ourselves.

When people close their eyes, some claim to see only blackness. In the Stanford tests blackness was not acceptable. Researchers claimed that impressions were bound to come through the blackness if the test subjects simply trusted their minds. Let's do a short exercise now that will illustrate this concept.

Close your eyes and imagine your bathroom at home. Perceive it clearly...create every detail in your mind. Ask yourself where the wash basin is located, or the toilet, the shower, the tub...what do the flooring and walls look like? Where were the towels and toiletries when you last saw the room? Close your eyes now and answer all these questions.

Good. Let's do just one more exercise before proceeding. Once again close your eyes and imagine an Indian riding on a horse. Now you've seen this image many times in books and movies, so draw upon these memories to create a vivid mental image of an Indian riding a horse. Create the environment in which the Indian is riding. Do this now for about 30 seconds.

You just perceived impressions. You made them up and they appeared in your mind. You allowed them to come in through the blackness. And this is the very least you should perceive in subjective explorations. You saw your bathroom with your inner eyes and you can see your past lives, or future potentials, or hidden information in exactly the same way...and probably with much more intensity. Some people may challenge the validity of the exercise, saying that the pictures of their bathrooms are being pulled from their memories. That's right. Of course that's what's happening. Everything is recorded in the memory banks of your subconscious mind—every thought, every action, every deed from this life or any previous lifetime. You absolutely have the power and ability to perceive them all in vivid detail. The key is to PERCEIVE. You don't have to SEE. Perception is simply a matter of knowing or thinking, and if you trust yourself it will work for you.

Probably the single most valuable receiving tip is: trust the very first thought that pops into your mind. This is what most psychics do. There is no little red light that goes on in a psychic's mind when he is receiving a psychic impression. He simply learns to trust that initial impression and let it unfold naturally. If he begins to analyze, intellectualize or question, he loses the impression. The same is true of impressions received in an altered state. So try not to question the process while you're receiving. Don't question whether you're perceiving valid impressions or making them up. You'll have plenty of time for that after the session.

One of the primary questions that comes up is: "How do I know that what I'm receiving is valid?" There is no simple answer. Rather than proving or disproving results, why not judge by the effect on your life? Does the information you received explain a situation, decreasing your anxiety about it? Does the awareness resolve an old problem? Does it make you feel better about yourself or someone else?

It's our belief that you wouldn't receive positive benefit if the information wasn't valid on some level.

Some people judge the validity of their altered state experience by researching the historical accuracy or through other forms of verification. If the impressions are prophetic and the prediction is fulfilled, there is little doubt about the validity of the information. In the final analysis, you are the judge of your own experience.

4

HOW TO USE YOUR TAPE

How Long Will It Take?

If you are using altered-state-of-consciousness techniques to accelerate positive change, there is no way to know just how long it will take. In some cases just a few sessions have been enough for a fully conditioned subject. Some need no conditioning at all. For most people change does take time, but once you're conditioned, if you use the tape every day, you should start noticing a difference within a very few weeks. Often it will depend upon how much negative reinforcement has to be overcome. If you have smoked two packs of cigarettes a day for 20 years, that is well over a quarter million times you have programmed the negative. That is quite a bit of programming to overcome. Be patient and continue to program your computer with the positive. It will work if you have the willpower to stick to it.

When people stop or find excuses such as "There wasn't enough time to do my hypnosis today," they really aren't ready to let go of their problems. It was more important to watch a TV show, talk on the phone or read a newspaper than to take 15 minutes for reprogramming their life. Everything in life is a value judgment and only you can decide what is most important. But don't excuse yourself due to lack of time and don't complain any more about your problem if you place it low on your importance list.

The people who stick to it get the results they are after, often rising above logic and baffling medical science.

If you are using the tapes to have mental experiences, the intensity of the experience will probably be dependent upon your willingness to trust the impressions you receive. Once you are conditioned you may receive immediately if you don't block yourself with judgment and criticism. In other cases, it may take many uses of the tape before you trust yourself enough to allow the mental experience to happen. If you do have trouble receiving, read and reread the "How to Receive Vivid Impressions" section of this book.

A fabulous hour-long tape is also available: "How to Be a Better Receiver," Tape P227— $3.95. It is an intense hour of Dick Sutphen conducting exercises and special hypnosis sessions designed to maximize your receiving effectiveness in situations such as regression, psychic input, Higher-Self impressions, etc. Everyone can receive well once they understand the various ways and experience special exercises. It is usually an individual's expectations that block the experiences. In an hour you can learn to receive with maximum effectiveness.

To put the understanding to immediate use, the last 15 minutes of the tape is a psychic exploration and short regression to demonstrate how you "can't help but receive when you know how."

How Many Things To Work On At Once

Once you've discovered how extremely effective hypnosis, meditation and sleep programming are, if you're like most people you'll probably want to begin to work on many changes at once. Try to refrain from doing this. You'll have more immediate results by concentrating your efforts on one or two primary directions until you have achieved your goals, then moving on to a new area of programming.

You can certainly work on several related things at one time. For example, you might choose to develop your concentration abilities at the same time you're using pre-race hypnosis right before a marathon and feet and leg relaxation hypnosis right after the race. One excellent combination would be to program your short- and long-term goals, plus health/healing suggestions. As your goals change you can also change your programming. Some people have reported using the tapes 10 times a day. That is entirely unnecessary. One daily session is adequate and two is the maximum suggested.

Short-term programming examples: Stop smoking, lose weight, stop biting nails, stop being sarcastic, remember people's names, develop concentration, expand creative ability, develop sports abilities, etc.

Long-term programming examples: Build a new home, make a major career change, own your own business, create financial success, develop a self-actualized perspective, create a new reality, etc.

When to Use Your Tape

Use your tape whenever it's convenient for you—ideally when you are wide awake and alert so you won't go off into a normal sleep. If you are using altered states to heighten creativity or some other similar activity, use the tape just before you plan to begin working. If you are using a physical activity tape, such as sports, plan to complete the session about a half hour prior to your activity or encounter. The reason for this is that you have been lying down totally relaxed, and it's best to get up and get your circulation going before you jump immediately into intense physical activity.

Body Position

Try to pick a time for hypnosis when you will not be interrupted and a place where it is quiet. You may sit in a comfortable chair or lie down in bed. If sitting, place both feet flat on the floor and your hands on your legs. If lying down, do not cross your legs. Weight can exaggerate during hypnosis. Place your arms at your sides.

The lying down position is best unless it causes you to fall into a normal sleep. Avoid using your tapes when you're very tired. The tape will condition your subconscious mind, and you don't want the tape to condition it to fall asleep when you go into an altered state. If you fall asleep two times while in the prone position, continue further sessions in a sitting position for a few days. If you don't do this your subconscious will quickly become programmed to fall asleep every time you use the tapes. The subconscious mind contains all of the memories of this life or any other life you've ever lived, but it has very little reasoning power. Thus it can easily be programmed contrary to your conscious desires, unless you know how to work with it. There is no danger whatsoever in falling asleep while using the tape; it is only the sleep habit pattern established in the subconscious that is to be avoided.

If you wear contact lenses and normally remove them when you go to sleep, take them out before working with your tape.

Deep Breathing

Yoga or meditative breathing should be used to relax your body and mind prior to entering an altered state. Take the position you will use during the session (sitting or lying down), set your tape player beside you so you need only reach over to press the play button. Now take a very deep breath...let it out slowly between slightly parted lips. When you think the breath is all the way out, pull your stomach in to push out even more air. Take at least five to 10 of these deep breaths before turning on the tape. *Note:* If you have a health problem or it is undesirable to use deep breathing for any reason, simply lie down and relax, breathing normally for several minutes before beginning hypnosis.

When you have completed deep breathing or relaxation, turn on your tape player and close your eyes. The tape will instruct you from that point.

Psychic Center

On the deep-level meditation tapes, you will be asked to raise one hand and use three fingers to apply gentle but firm pressure upon the primary psychic acupressure point, located just below the center of your rib cage and above your stomach. After about a minute you'll return your arm to your side.

Body Relaxation

On most tapes, the first instructions are for body relaxation. You are instructed to "feel" the relaxing power coming into the toes of your feet...and then proceed to relax your entire body. You can help by "playing the role." Actually "feel" your feet, legs, arms, etc. relaxing.

After the body relaxation is the actual induction. Keep your full attention on the voice on the tape and use your active imagination very, very vividly to visualize yourself in a situation as described on the tape. If the instructions are to go "down, down, deeper, deeper," create a pleasing mental picture of yourself walking down a forest path or the steps of a building, climbing down rocks to the sea or whatever appeals to you. Do this with as much imagination as possible. If the instructions are to go "up, up, higher, higher," mentally create a situation in which you are floating up on a cloud, or riding an escalator. Your ability to fantasize at this point is actually the most important part of the induction. You put yourself into an altered state of consciousness by becoming an active mental participant.

On most tapes, once the induction is complete you are given a self-release suggestion so you are conditioned, no matter how deep, to awaken yourself at anytime you desire.

At this point the programming content of the tape begins. Of course, the content varies with each tape.

Extending Your Time

Once you have listened to a tape you will be aware of the length of time allowed for you to receive input or to create positive affirmations. If you want to extend this time you can go into meditation with your finger on the pause button of your tape player. When you come to the section of the tape you wish to extend, simply press the pause button. This action will not bring you out of an altered state.

Another option is to have someone else in the room with you running the tape recorder according to your preference.

Retaining Your Impressions

Subjective impressions received in altered states are often like dreams in that they quickly fade. For this reason, you might want to have a pencil and paper beside you when you awaken so you may quickly write down the highlights of the session. You might speak into a tape recorder so you can keep your eyes closed and verbally commit your experiences to tape while they are still fresh in your mind.

Many people use a second tape recorder, leaving it on while the first tape plays. They then speak up and verbalize their impressions as they are occuring. The result is the same as a directed past-life regression. Speaking up will not bring you out of an altered state; in fact, it's a great solution for keeping you focused on the input if you have a tendency to drift off or fall asleep.

The Tripping Problem

Once you are fully conditioned, you may sometimes experience going into an altered state and not remembering anything until you awaken. If you are opening your eyes on the count of five, you are not going off into a natural sleep. You may actually be too good a subject and are doing what we call "tripping" or drifting in and out. There are several solutions.

First, try sitting up against a wall or in a chair while exploring in an altered state. You won't be quite as comfortable, but this may help to hold you from tripping.

If you are simply going too deep, don't do any deep breathing before the induction. Once you have become conditioned, you may also want to shorten the induction by imagining a wave of relaxation moving from your toes to your head. Another technique is to keep yourself fully conscious during the initial part of the induction. On the hypnosis tapes, begin participating at the second seven count down; on all other tapes begin to participate about halfway through the induction. If you want to use this technique simply advance your tape player to the beginning of the second count down before beginning your own relaxation.

Other techniques that may help if you're falling asleep or tripping out: if you're familiar with yoga postures, do about 15 minutes of them before going into trance. A real natural "upper" that will help keep you alert is vitamin E and honey. Combine the two about 20 minutes before you go into trance. This will give you a "speed" effect that will last about four hours. The yoga, honey and E combined will have you so alert you won't believe it. The honey instantly puts sugar into your system and the E extends the oxygen. Don't use this technique if you've been drinking alcohol for it will work in reverse—as a downer. If you wish to avoid honey, take two super-high quality B Complex vitamins at least 30 minutes before you go into hypnosis. B's are fatigue fighters and will help you remain alert.

Spinning or Swaying

A small percentage of people sometimes experience the feeling of spinning or swaying while in an altered state—especially towards the completion of the induction. There is nothing to fear: simply give yourself the strong command: "stabilize!" You are in control and you can stop the effect.

Headache

Very rarely someone will awaken with a headache that feels like a tight band around the forehead. Although somewhat uncomfortable, it is not a matter for concern and will usually disappear within 30 minutes. The ache can be the result of anxiety about the altered-state experience, but those involved in metaphysical investigation feel it is the result of "third eye" activity and indicates the awakening of psychic abilities. Using altered state techniques often results in an expansion of extrasensory perception even when the subject isn't trying to be psychic.

5

PROGRAMMING

To begin, let's discuss the subconscious mind and how it works. It is the subconscious that you work with in positive programming and reprogramming. Think of yourself as a mind, for this in reality is what you are. You do not have a mind, you are a mind. You are the sum total of all your experiences from the time of your birth up until right now. If you believe in reincarnation, that would also include all of the experiences of your past lives.

Regardless of your belief system, it is these past experiences which represent all of your programming and all of these memories are retained in the subconscious memory banks. It is your subconscious mind that has made you what you are today. Your talents and abilities, problems and afflictions are the result of the intuitive guidance of the subconscious. It has been directing you and it will continue to direct you...and often in opposition to your conscious desires.

Why? Because the subconscious has little or no reasoning power. It is simply operating like a computer...**functioning as a result of programming. As a medical fact: The subconscious creates only according to programming.** It will help to bring into actuality the reality for which it is programmed. If your subconscious were to receive no new programming, it would continue to operate on all of the prior programming of your past. This, of course, cannot happen, for you are constantly feeding new programming or data into your subconscious mind—your computer. Every thought programs the computer. Thus if you are thinking more negatively than positively, you are programming your computer the wrong way. You create your own reality or karma, as some people call it, with your thoughts. **You create your own reality with your own thoughts.**

Many of you have no idea how negatively you think. If you climb out of bed cursing the alarm clock, grouch through breakfast and then think negative thoughts about the rain, the heavy traffic on the freeway, how much you dislike your job and on and on through the day...you are literally creating a negative reality for yourself. With all that negative programming of your computer, how could it do anything but create the programmed negative result: more negativity.

Your thoughts from the past have created your today. If you don't like your today and desire to change it, it is simply time to change the programming of your subconscious computer. **What mind has created, mind can change.**

The following information, based on recently proposed physics theories and brain/mind research, will help you to better understand the importance of the programming of your mind.

New Research Explains Why Altered States Make Your Life Work Better

---

You lie quietly in total darkness. From far away you hear your house murmuring through its pipes and heating system, creaking with the wind outside. It is not a part of your world. Neither is your body, which now seems weightless and unimportant. Your entire reality is within your mind. You are in an altered state of consciousness.

A single question winds through your mind, repeating itself and twisting this way and that so you can perceive it from all sides. "Why am I so afraid of success?"

You've asked yourself the same question many times while in full Beta consciousness, with no results. None of the logical, rational answers seem to fit. You know you have the talent, background and expertise to be succesful. Yet the fact remains that you are not.

"Why am I so afraid of success?" you ask yourself again. You feel yourself slipping into an even deeper trance, and realize that soon you will be unable to focus your attention on the question at hand.

"Why am I so afraid of success?" In a blinding flash, you have the answer: you are afraid to be successful because you need the challenge and

> excitement of reaching ever further. Subconsciously you are afraid that if you allow yourself to be successful, you will have nothing left to want, to strive for, to create aliveness in your life.
>
> There's a peculiar clarity that comes with these insights. An internal, instinctive response which, if verbalized, would be: "But of course." There is no doubt in your mind that you have uncovered your block to success, and that you will now be able to rise above it.

So you come back to full consciousness feeling elated and ready to tackle the world. And in that part of your life you can. Of course, your relationship still isn't working very well and you still eat more than you should, but you can't expect miracles. Right?

Wrong.

According to leading-edge physics research, the use of altered states of consciousness can lead to a transformation of nearly every part of your life.

Physics chemist Ilya Prigogine recently proposed a theory that earned him a Nobel prize. The theory—already confirmed by experiments—is called "the theory of dissipative structures." It has solved the mystery of why the use of altered states can result in life-changing insights, new behavior patterns and the relief of life-long phobias or ailments. Here's how the theory works as applied to real people:

First, you must understand that human beings are structures. The structure of your body is composed of bone, muscle, ligaments. **Your brain, however, is given structure by the thoughts and memories that dictate your actions. It is the programming of your brain that provides its structure.**

Now, Prigogine's theory states that complex structures (such as the human brain) require an enormous, and consistent, flow of energy to maintain their structure.

In the brain, that energy is measured as brain-wave levels on an EEG machine. The up-and-down pattern of brain-wave levels reflects a fluctuation of energy to the brain. The larger the fluctuation of brain-wave levels, the larger the fluctuation of energy.

In full, or beta, consciousness, your brain-wave levels would show up on an EEG graph as small, rapid, up-and-down lines (see Diagram 1 below). There is little fluctuation in the level of energy.

(Diagram 1. Beta brain-wave levels maintain a fairly constant flow of energy through the brain.

However, when you alter your state of consciousness through the use of hypnosis, meditation, relaxation, etc., your brain-wave levels shift to alpha and theta (see Diagram 2 below). In these altered states there is a lot of fluctuation in the level of energy.

(Diagram 2. Alpha and Theta brain-wave levels create large fluctuations of energy through the brain.)

According to Prigogine's theory, small fluctuations of energy (such as beta rhythms) are suppressed by the brain so it stays essentially the same.

That's why changes suggested to a conscious mind usually have little effect. The message is suppressed by all the existing programming.

However, says Prigogine, large fluctuations of energy (such as alpha and theta rhythms) can cause the structure to break apart and reorganize itself into an even more complex and higher form.

That's why suggestions given to an individual exploring in the alpha and theta brain-wave levels are so effective in creating change. The new suggestion, dropped into the uneven alpha rhythms like a pebble in a pond, creates a ripple effect that tears apart old programming and creates new behaviors and viewpoints.

What's happened is that your brain has broken down its old programming and reorganized it into new, more complex, and usually more meaningful forms.

When this shift occurs, you may become aware of information about your life and goals that the old structure of memories and programming kept hidden from you (such as in the first example); you may experience a sudden, powerful insight into an old unsolvable problem; you may release yourself from the programming effects of a traumatic memory.

Afterwards, it's often difficult to describe the experience. You don't really know why you so firmly believe this is it; this is the block you've been seeking.

Sometimes the explanations aren't even logical.

For example, the case study of Nancy, who had a terrible case of atopic dermatitis, a chronic skin condition that leaves your skin with itchy, open sores. Practically everything aggravated her condition: too much sun, too much sweat, too much soap. Nancy had borne the disease since childhood, and doctors told her she could expect to bear it for the rest of her life.

As a last resort, she went into an altered state of consciousness and asked herself the question, "Why am I creating this skin condition for myself?" The answer she received was: "One of the primary lessons you want to learn in this life is to be sensitive and open to other people. You choose to have a super-sensitive exterior—your skin—to remind you to reveal your even more sensitive interior—your true Self. When you do this, your skin condition will disappear."

After coming back to full consciousness, Nancy stated that, although logically she shouldn't believe a word of it, for some reason it felt right. She admitted that she never allowed her true Self to be exposed to others for fear of rejection.

Several months later Nancy wrote to say that, since that session, she had followed her own advice and begun revealing herself to others who became her friends. Her skin condition had cleared up entirely.

By searching for a solution to an "unsolvable" problem while in an altered state of consciousness, Nancy had caused her mind to reorganize all of its beliefs about the nature and purpose of her disease. Perhaps in the process of creating new and more complex structures, her mind recognized and used its own ability to heal the body. Or it could be that the skin disease was psychosomatically induced, and thus when her mind reorganized its structures the purpose for the illness was eliminated.

Another aspect of the transformation that can occur through the use of altered states is the exploration and release of an old, negative memory.

In a recent Sutphen seminar, a woman named Terri stood up to share a childhood experience that still moved her to tears. She couldn't do it. The memory was so traumatic for her that she couldn't bring herself to speak of it. When Dick saw this he decided to do a one-to-one regression.

After inducing hypnosis, Dick asked Terri to go back to the event in her childhood that was responsible for her painful memories. She began to describe a time when she was about four years old. She was just becoming aware of her body and, as most children do, she began to explore and touch herself. One day her mother discovered her and was so enraged that she beat her daughter badly and called her names. Terri was so traumatized by the childhood experience that as an adult she still felt enormous guilt and shame about it. It was ruining her sex life.

Throughout the altered-state session Terri cried, but after completely reliving the experience something interesting happened. Her body language, which before had reflected an insecure and childish woman, became strong and adult. Her voice was more confident. When she opened her eyes she looked at Dick and said, "I can't believe I took that experience so seriously. Obviously it was my mom's hangup, not mine."

By reexperiencing the traumatic memory in an altered state of consciousness, Terri had triggered a transformation of that memory. Its power over her was lost in the process.

And that's only the beginning of what is possible to achieve through an altered state of consciousness. According to Prigogine's theory, there's also an added bonus: **each**

transformation makes the next one likelier!

You see, every time you trigger a collapse of memory or data structures, your brain reorganizes its structures into even more complex, more highly organized forms.

As a result, it requires even more energy to maintain those structures. And those structures are even more vulnerable to fluctuations of energy. **Basically, the more complex a structure is, the more unstable it is, and the easier it is to trigger the next transformation.**

If you carry this concept to its logical conclusion, every time you **successfully** use an altered state of consciousness to achieve insights or solve a problem, you **increase** your chance of success the next time.

That's when the miracles begin.

Reality/Fantasy Tests

Studies in the brain/research and university laboratories have also brought to light another aspect of the ease with which the subconscious mind is programmed. One of the initial tests was the recording of actual brain-wave patterns under specific conditions.

The purpose of the study was to determine if the mind can distinguish fantasy from reality, if it accepts as programming only that which is real, or if it also accepts that which isn't.

In this test, subjects were placed in a room and wired up to an EEG machine. The researchers then created several situations to see what effect the events had on the brain-wave patterns of the subjects.

For example, someone ran into the test subject's room and fired a gun. Another person performed a dance. A dog barked. A color was projected on a screen, and many other situations were created. As each situation occurred the test subject's brain-wave patterns were monitored and marked as to which situation created which brain-wave patterns. For example, a segment of brain-wave patterns might be marked, "Dog barked here."

The next step was to have the same subjects mentally recreate the situations as they were described by the researchers. One example is, "I now want you to imagine you are watching a woman doing a dance. See it in your mind, fantasize it, conceive it with as much imagination as possible...All right. I now want you to imagine a dog barking."

While the subject was concentrating on these imagined situations, his brain-wave patterns were again being recorded. The test results showed that the exact same brain-wave patterns were evident when the subject actually experienced the situation, as when the subject only imagined the situation.

The brain waves were identical, so the computer part of the brain was obviously incapable of telling the real from the imagined.

Mental Programming Tests

Another supportive series of tests were

conducted by the University of Chicago. These and many similar tests show how your subconscious computer actually creates the reality for which it is programmed. Three test groups of students took part in a mental programming experiment based upon shooting basketball. All the participating students were tested as to their individual basket-shooting ability and the results were recorded.

Group One was told, "Don't play any basketball for a month. In fact, just forget about basketball for the entire month."

Group Two was told, "You are each to practice shooting baskets for one full hour a day, every day of the month."

Group Three was told, "You are to spend one hour a day imagining you are successfully shooting baskets. Do this every day for a month. Imagine or fantasize yourself at being successful shooting baskets. See every detail of your accomplishments in your mind."

One month later the three groups were again tested as to their ability to shoot baskets. Group One, who hadn't played basketball for a month, tested exactly the same as they did the first time. Group Two, who had been practicing a full hour a day for a month, tested 24 percent improved in their basket-shooting ability. Group Three, who had only imagined they were successfully shooting baskets for an hour a day, tested 23 percent improved in their actual basket-shooting ability—only one percentage point less than the group that had actually been practicing!

Obviously, the group that only imagined successfully programmed their subconscious computers to perform almost as effectively as those who actually practiced. The subconscious can be fooled. It can be tricked. It can be programmed and you simply have to know how to become a programmer. It is my belief that if the group that only imagined had used altered-state-of-consciousness programming techniques, they might even have exceeded the results of the group that actually practiced.

Using the Knowledge

This is not to say that the use of altered states can replace training and practice in sports that require a high level of physical fitness and skill, like karate, running, tennis, etc. You know that to be good you must develop your physical and mental ability through practice.

What has been proven, however, is that the use of altered states of consciousness can be an invaluable tool for programming your subconscious mind to work for you, rather than against you.

What Can You Expect?

First, you can expect altered-state-of-consciousness programming to accelerate your comprehension and learning skills. Your self-assurance and concentration abilities will increase, as will your intuition. Another major benefit is added willpower and motivation.

Some of the specific results you can expect in different areas of your life are:

Career: Altered states provide the ultimate methodology for understanding and releasing your self-imposed blocks to success. It will give you the confidence and charisma you need to improve your sales figures, pitch your ideas successfully, become a more effective manager. Altered-state-of-consciousness programming can shorten the time it takes you to learn a new skill or earn your next promotion.

Social Skills: If remembering people's names is important to you, you can learn how to do it through altered states. It can also help you to increase your confidence when meeting people, improve your public speaking skills, and increase your intuitive ability to understand people.

Bad Habits: One of the most common and effective uses of altered-state programming is to overcome bad habits such as smoking and overeating. Other bad habits that yield to programming are nail biting, excessive drinking and nervous habits.

Self-Improvement: Altered states are highly effective in improving your own inner reality. It is the ultimate technique for eliminating tension and stress, for creating a positive attitude, or for increasing peace of mind.

Psychic/Higher Mind: And finally, altered states are the most effective known methodology for encouraging psychic activity and Higher-Self awareness. ESP experiments, psychic input receiving or any form of vision induction are extremely successful, for the process can be directed to a consciously alert individual who is exploring in the alpha, theta and delta brain-wave levels. The resulting "communications" or "impressions" are usually quite vivid.

Most people who inquire about altered states are interested in one of the following: **1. Overcoming a problem. 2. Accomplishing an objective. 3. Having an experience.** Altered states of consciousness are certainly no magic wand, but when used correctly it can give you an edge. It can provide you with a running start and help you to open all the necessary doors as you proceed towards your goals.

6
OTHER TAPES

NEW AGE MUSIC TAPES

The music of Higher Consciousness—created to calm the mind and nourish the soul. New Age music is usually written in a pantanic scale. This means it lacks the tension and resolve evident in all other music except Eastern and oriental pieces. By combining Western musical tastes with this ancient concept, Americans are realizing delightful psychological/physical benefits along with their listening enjoyment.

New Age music tapes are produced one of two ways:

"Inner Harmony"

1. A soothing, inspiring musical environment is created and sustained for the full length of the tape. There are no dramatic shifts or changes, although the music evolves in rich and subtle ways. An example would be the mix of UPPER ASTRAL SUITE (Tape AM105). It combines many instruments, synthesizer, delicate chimes and bells, trickling water in a creek and ethereal voices humming soothing waves of sound patterns. The sustained, yet subtly changing sounds support and enhance your altered-state-of-consciousness session without the slightest distraction. It also serves as non-distracting, yet uplifting background music for any appropriate activity.

"Progressive"

2. Music created to inspire visualizations.

It is structured with a theme in mind that invokes feelings, emotions, or impressions.

It is believed in ancient metaphysical cultures, such as Egypt and Atlantis, that musical keynotes in certain combinations were used to stimulate, awaken and balance the chakra energy centers that lead to expanded awareness. This is the goal of today's New Age musicians— to produce an inner harmony on the part of the listener. The ancient sacred writings of India, the Vedas, tell of the incarnation of gifted musicians at the beginning of each new age. They are said to herald the transition and assist man in his awareness of the mysteries.

Recent scientific studies of the relationship of music to the illnesses of the body and mind have yielded some startling results. New Age music proved far superior even to classical music in producing dramatic change in consciousness as measured upon scientific instruments. Listeners transcended levels of awareness, achieving serene and relaxed states.

Other tests have shown that New Age music often deepens breathing patterns, which help regulate blood pressure and relax the body and mind. Thus many hospitals, schools and institutions use this music to create a more nourishing, soothing, healing environment.

We suggest that you use New Age music as background or foreground music with any of the following activities:
• Social gatherings
• Relaxed time alone

- Altered-state-of-consciousness sessions such as self-hypnosis, meditation or creative daydreaming
- Massage
- Creative activities such as writing, drawing, painting, etc.
- Working
- Lovemaking
- Astral projection
- Behind spoken meditations in yoga classes, seminars or New Age groups

MEDITATIVE SOUND EFFECT TAPES

These sound effects tapes were all created in the actual environment. Examples: Ocean waves hitting the beach accompanied by the sound of seagulls calling. Forest sounds of a light breeze in the trees, birds and crickets. Gentle rain in the mountains with the far away sound of thunder. Creek in the forest with the sounds of birds and crickets. Rain on a Malibu redwood deck, plus a Solari wind chime, an occasional seagull and ocean waves hitting the beach.

These tape are designed to create a soothing, sustained environment. They are ideal as background sounds for altered-state-of-consciousness work or to create a warm environment for any activity. Many people prefer these tapes to using white noise effects to block out distracting environmental noises.

SYMBOL THERAPY TAPES

Symbol Therapy is "psychosynthesis" for psychological and spiritual growth. Pyschosynthesis is a psychology of self-actualization and a philosophy in the art of living. It is visualization meditation to change your life.

In the Symbol Therapy tapes, Dick Sutphen relaxes your body and then directs you through a beautiful, superconscious visualization which can have life-changing effects if listened to regularly. Each tape uses different, soothing New Age music in the background and is approximately 20 minutes long, with the same program repeated on both sides.

"Psychosynthesis" is explained in the book, **What We May Be,** by Piero Ferrucci, in which the value of symbolic visualizations/meditations is stressed: "When we visualize them, symbols can have a profoundly transformative impact on our psyche. They can be used to undo past conditioning and to establish new and more desirable energy patterns within us."

Ferrucci worked for years with the famous Italian psychiatrist, Robert Assagioli, whose psychosynthesis techniques and principles have assisted thousands to find self-realization.

Psychosynthesis uses powerful and effective superconscious visualizations. In his book, Ferrucci explains that we become intuitively receptive to the essence of a symbolic image by holding it steadily in our mind, and that such contemplation of an image may lead to an **identification** with it. Subconscious identification with a positive symbol can have positive, rejuvenative mental/physical effects.

Of the symbolic visualizations, Ferrucci says, "They serve to structure and direct certain unconscious energies. We experience a lot of nervous stimulation but end up by doing nothing. This free-floating energy can often cause exaggerated emotional ups and downs, produce dispersion and a sense of meaninglessness, or explode into a sudden outburst of anger. It can also manifest itself as some general feeling that something needs to be done — but we know not what — and in many other unproductive and disturbing ways.

"The unconscious, and particularly this disorganized, chaotic part of it, needs to be coached. It needs to have a rhythm and a direction communicated to it. Evocative symbols can greatly help in this task, because they tend to focalize free-floating psychological energy without repressing it."

By using Symbol Therapy, you can focus your energy to change your reality.

7

IDEAS TO MAKE THE TAPES MORE FUN AND EFFECTIVE

1.
Mental Experiences Tapes: Group Experiments

Once you are familiar with the tapes you may want to conduct a group session with your friends or family. Explain to the group what you have learned from the sessions and about your own experiences in the sessions. Read them the pertinent sections of this manual or discuss it in detail to ensure understanding. Then have everyone stretch out on the floor and begin to direct them through the deep breathing exercises. Once this is completed, turn on your tape player. After the tape is over, start a conversation about what everyone experienced.

If any of your participants are hard to awaken or go off into a natural sleep, you can simply awaken them normally. If that doesn't work, raise one of their eyelids and blow on it gently They will quickly awaken.

2.
Combining Programs

If you only want to do one altered-state session per day, yet you desire to work on a couple of different concepts at the same time, you may rerecord the appropriate portions of two or more tapes onto one longer tape. Please keep in mind the information expressed earlier in the book regarding how much to work on at one time.

You can also program your sleep tapes to tackle more than one goal in a session. For example, you may want to combine goal sleep programming (from the Goal Achievement Album C815) and eliminate worry sleep programming (A109). You'll need two tape recorders—one to play the tapes and another to record. Use C-120 hour per side blank cassettes and record all of the goal tape except the final instruction to go off into a deep natural sleep. Then advance the eliminate worry tape past the initial relaxation until the actual programming begins. At that point record the rest of that tape onto the last half of the hour tape.

Our research indicates that sleep programming is not effective for more than an hour per session. Within that period you attain a sleep level so deep that your conscious mind is not connected enough to accept programming.

Please do this rerecording for your own use only. Duplication for others is a violation of legal and karmic law.

3.
Self-Hypnosis Without The Tape

Once you have worked with a prerecorded hypnosis tape you have the basis of self-hypnosis without the tape. All you have to do is duplicate the tape induction...it doesn't have to be word for word. Simply use your breathing technique, then relax your body by mentally giving yourself the suggestions on the tape. You will soon be able to relax your body much more quickly than a standard relaxation. Count yourself down and begin to repeat the positive suggestions you have created for yourself. Create a mental movie that supports the achievement of your goal. When that's completed, count yourself back up as on the tape. It's that simple, and the more you work at it the more effective it will become.

You may go into a deeper trance when you listen to the recorded voice, but this is to be expected. It does not mean that doing it yourself is ineffective...it will be extremely effective. Just keep at it.

You can use the suggestion time to program yourself for an instant response if you desire. For example, you could condition yourself to react to the phrase, "sleep now and rest," and in so doing you can develop the ability to immediately go into trance, quickly give yourself a suggestion and then awaken to full beta consciousness on the count of five. The entire process might take a minute or two.

If you desire to develop this ability, try phrasing your suggestion this way: "I am now conditioning my mind to respond to a key phrase for instant response. When I breathe deeply three times and say the words 'sleep now and rest,' I will instantly go into a deep hypnotic sleep. 'Sleep now and rest' is the key phrase for instant conditioned response (repeat and repeat).

Hypnosis is not a magic wand, but it is

certainly an effective tool. You are not a robot, so you can't expect instant programming miracles. However, you can, once conditioned, expect results that far exceed what you could have achieved without hypnosis. Remember though, you must experience self-hypnosis on a consistent basis to be a conditioned subject. You will get out of self-hypnosis what you put into it. Once you have developed the learned technique you can use it for the rest of your life, if you continue to retain your ability by using it at least two or three times a week.

4
Your Environment

Certain conditions are more conducive than others to an altered state. An overly warm room is much better than a cool one. Darkness is better for most people to visualize effectively, so if you don't have a dark room use one of the sleep masks available in any drugstore for a few dollars. If your environment is noisy use earphones plugged into your tape player. If it is extremely noisy you may also want to play another tape at the same time to block the noise; sound effects tapes of rain or the ocean are good. New Age music can also be of value here. If you don't have two players maybe you can borrow one long enough to rerecord the background sound on one channel and the altered-state program on the other. Then you'll hear both on the same tape when you play it back.

5.
Eye Opening Experiment

If you are using a program, such as one of the many psychic and metaphysical tapes in our line, to generate mental experiences, you may want to attempt an interesting experiment. Those who are in a **very deep** altered-state-of-consciousness can open their eyes without affecting their trance level. If you feel you are such a subject, go into trance with a pen and paper by your side and open your eyes to draw a picture of what you are mentally observing. Then close them again and continue to participate with the tape as you normally would.

6.
Touching Others In Trance

Touching others in an altered state can sometimes create a psychic connection. As an experimental example, let's assume you'd like to explore the concept of a shared past lifetime. Before using one of the regression tapes, make the conscious decision to seek a previous lifetime you may have shared. Communicate this idea clearly to your subconscious mind. Then go into the altered state holding hands. Once the tape is complete verbally compare regression results. To be more objective, each of you write down your experiences and then share them.

7.
Physical Considerations

Some esoterically-oriented people feel that they attain superior results with the metaphysical tapes by following one or more of these occult principles:

- Go into an altered state lying down with your head pointing north.
- Remove all metal jewelry.
- Remove all clothing.
- Surround yourself with three lighted white candles.

8.
Chakra Connection

An intense psychic connection can be established between two people by conducting a chakra visualization exercise prior to using the metaphysical tapes to have mental experiences. The chakras are the seven centers of the etheric body. Lie down side by side, hold hands and do your deep breathing. Then one of you should verbally direct the connection:

"We are now going to visualize a connection taking place between the crown chakras at the top of our heads. Visualize an intense violet light emanating from the top of your head...see it in your mind...it is arching out, up, across and over to connect with my head. Make it real with the unlimited power of your mind...it is an intense, shimmering, irridescent violet light connecting our crown chakras. (pause) The connection is now complete, and it is time to visualize the next chakras:

Both people participate in the visualization, and the process continues with the next three chakras:

Third-eye, brow chakra (in center of forehead)—blue violet in color.

Throat chakra—silver blue in color.

Heart chakra—golden in color.

With practice many couples are capable of perceiving identical impressions, emotions or regression experiences in an altered state of consciousness.

9.
Group Games

Groups of people have successfully attempted many techniques while exploring subjective tape experiments. Some examples:

1. With a group of three to 12, lie down and form a wagon wheel with all heads touching and bodies fanning out to form the spokes of the wheel. Compare experience notes after the session. The **Frequency Switch (S117)** tape is especially good for this one, but any metaphysical tape will work.

2. Using the Chakra Connection technique switch couples after each session to see if you are more psychically attuned to one person than another.

3. Using the **Mind Projection (S118)** tape, send one person to a destination of their choice

within 20 minutes driving time. No one in the group is to know the destination. The entire group then goes into an altered state and projects to that person. Upon awakening the group compares notes about the experience and verifies the projections when the person returns.

4. Group human potential processing can easily result from the use of many of our tapes (See complete catalog). Share ideas and assist each other to ask the right questions to rise above current conflicts and problems.

5. For the esoterically-oriented group, the **Higher Self Explorations Album (C811)** offers ideal group explorations such as "Verbal Channeling," "Automatic Writing," and "Telepathic Contact On A Superconscious Level."

8

PROBLEMS

Q. I keep falling asleep and I don't wake up until the tape is over. Am I still getting the benefit of the tape's programming?

A. You may be receiving some benefit, yet you are unable to participate in the mental movies and other techniques requiring your involvement. If you are going off into a natural sleep you probably won't awaken upon being instructed to do so. If that is the situation use a sitting position and do not use the tapes when you are tired. If, however, you are awakening on the count of five, but do not remember anything that has happened, you are "tripping." This means you are too well-conditioned and need to reread the **"Tripping Problem"** portion of this manual.

Q. The tape instructs me to visualize, but I can't seem to do it. I've never been able to visualize. What can I do?

A. A very small percentage of people have this problem, or they think they do because they have unrealistic expectations about what's involved in visualization. Once again try the exercise described earlier in this book: Close your eyes and "remember" your bathroom at home. Remember the placement of the washbasin, tub or shower, toilet...the towel racks...the floor covering...the wall color and texture. Remember everything about your bathroom. Do it now before proceeding.

Alright, in your "remembering" of your bathroom you were probably visualizing without realizing it. With your innner eyes you were capable of perceiving your bathroom. Sexual fantasy is another area of internal visualization. It you've ever imagined a situation during masturbation or sex you are quite capable of visualization. You've just been expecting it to be something other than what it is.

If you are truly unable to visualize in any way, and it is important to you to develop this ability, use the following technique:

Turn on a TV and watch it for two minutes. Then turn it off and attempt to mentally remember everything you've just seen. Repeat the process over and over. Once you can do this successfully, move on to working with a novel. Read a page and then put it down and attempt to form mental pictures of the scene.

Q. I'm using the past-life regression tapes (or any of the subjective experience tapes), and I'm not receiving any impressions.

A. It is impossible not to receive impressions if you know how. Your expectations about how you "should" receive the impressions are blocking your experience. Reread the "How To Receive Vivid Impressions" portion of this manual. If you still feel you need additional help, you can order a special tape we sell: "How To Be A Better Receiver In Hypnosis" by Dick Sutphen, Tape HP042—$12.50. It is an intense two hours of exercises and instructions that communicate everything that can be said on how to receive. It is impossible to use this tape and not receive, as Dick demonstrates with the psychic explorations and a short regression. Send with $2.50 postage/handling to Valley of the Sun, Box 38, Malibu, California 90265.

Q. The key trigger words on my tape don't seem to be having any effect when I use them in my daily life.

A. The post-hypnotic suggestion trigger words become a little more effective each time you use your tape and each time you use the technique. Are you following the instructions to use the tape every day? This is critically important. Are you then using the technique on a regular basis? The tapes are not magic, but if you use them exactly as instructed they will continue to become more and more effective over the weeks, months and years. Do not be impatient and expect miracles. Continued use over a period of time will manifest fantastic results.

Q. Whenever I start to go into an altered state I have to itch or go to the bathroom. I get so distracted I can't pay attention to the tape.

A. Your subconscious mind is probably threatened about what you may discover and creates the distractions as an "avoid." Make sure you go to the bathroom first and then if you itch, scratch, but continue listening to the tape. Do not allow your subconscious mind to override your conscious desires. Continue with the process. Eventually your subconscious will get used to the techniques and stop creating the distractions.

Q. Instead of receiving impressions according to the instructions, I receive all kinds of unrelated impressions. In a past-life regression I received glimpses of five lifetimes during the time I was supposed to be dwelling upon one situation in one life.

A. This is quite common when you first begin to work in an altered state. Your subconscious appears to want to release a lot of information and is afraid it isn't going to get a chance, so it pours it all out at once. After you work with the tapes for a few weeks at most, this will stop and it will be easy to remain focused upon the desired individual input.

9

VALLEY OF THE SUN TAPES

Subliminals

The power of subconscious suggestion has an amazing effect on conscious behavior. Articles in **Time** magazine, **Omni, The Wall Street Journal** and **Brain-Mind Bulletin** illustrate how businesses are effectively brainwashing the public with messages piped through public address systems, flashed on movie screens, etc.

For instance, a **Wall Street Journal** story told the result of subliminals being used in a New Orleans supermarket. The message behind the music is, "If I steal, I will go to jail." The store owner claimed that before using subliminals, pilferage cost him $50,000 every six months. It dropped to $13,000 and cashier shortages went from $125 weekly to under $10! (This drastic difference is probably due to the fact that cashiers are continually exposed to the messages.)

While the ethics of using subliminals on the unsuspecting is questionable, (For more on how the public is being brainwashed, see "The Battle For Your Mind" tape, L802–$9.98.) programming yourself with subliminals is an easy and effective way to produce positive behaviors and eliminate negative ones, such as smoking, worry, etc.

The techniques used on the subliminals are not easily explained, but the voices are psychoacoustically modified and synthesized so that the subliminal voice is projected in the same chord and frequency as the music, changing and giving it the effect of being part of the music. If you were to listen very carefully, you might perceive what appears to be background chanting. Most people hear nothing but the music. If the affirmations were listened to separately from the rest of the music, it would sound like Gregorian chanting.

The result is fantastic. Subliminal messages are hidden behind beautiful, relaxing New Age music. There is no hypnosis or meditation. No concentration is required. You simply turn on the music, which you hear consciously, while your subconscious perceives the desired messages without resistance from your conscious mind.

Because you don't have to be in an altered state of consciousness for subliminals to be effective, you can listen to them at any time—while working, driving, relaxing, exercising, whenever! It's the easiest and most effective way to change your life!

*A small daily
investment
of working
with your mind
will yield
fabulous
returns.*

The following is a list of the Valley of the Sun tape line. Several new items are added quarterly, so write for a free copy of our current catalog: Valley of the Sun Publishing, Box 38, Malibu, CA 90265. Phone 818/889-1575.

2-TAPE ALBUMS WITH BOOK

C802	ASSERTIVENESS TRAINING
C814	ASTRAL PROJECTION
C838	ATTRACTING RIGHT LOVE RELATIONSHIP
C819	BUSHIDO
C842	CIRCLE OF LIGHT
C848	CRYSTAL AWARENESS
C847	CRYSTAL ENHANCEMENT
C823	ERASING SOMEONE FROM YOUR MIND
C824	GETTING BY ON 4 HOURS SLEEP
C815	GOAL ACHIEVEMENT
C830	HEALING ACCELERATION
C831	HIDDEN PERSUADERS
C820	HIGH ENERGY & ENTHUSIASM
C816	HIGH PERFORMANCE
C825	METAPHYSICAL AWARENESS
C821	NO EFFORT WEIGHT LOSS
AX901	PAST-LIFE THERAPY
C822	PERSONAL POWER
C837	RAPIDLY DEVELOP PSYCHIC ABILITY
C852	SELF-CREATION GUIDED MEDITATIONS
C835	SOULMATES
C827	SPIRITUAL POWER AFFIRMATIONS
C834	STOP SMOKING
C849	TEOTIHUACAN: REINCARNATION OF 25,000
C839	ULTRA-MONETARY SUCCESS
C844	UNCONDITIONAL LOVE
C833	VIEWING PAST LIVES
C836	WHY ARE YOU HERE?
C843	YOUR INNER TEMPLE

5-TAPE MIND POWER PROGRAMMING AUDIO/VIDEO ALBUMS

PK104 ACCELERATED LEARNING
PK107 ATTRACT LOVE
PK109 TENNIS PROGRAMMING
PK105 INCREDIBLE SELF-CONFIDENCE
PK103 LOSE WEIGHT NOW
PK106 SELF-HEALING
PK108 STOP SMOKING FOREVER
PK102 ULTRA-MONETARY SUCCESS
PK101 UN-STRESS

RX17 DIGITAL-HOLOPHONIC™ TAPES WITH SUBLIMINALS

RX103 A CALM AND PEACEFUL MIND
RX105 ACCELERATED LEARNING
RX131 A GREAT MEMORY
RX209 AKASHIC RECORDS
RX212 ALTERED-STATE SOUND
RX203 ASTRAL PROJECTION
RX102 ATTRACTING PERFECT LOVE
RX204 AUTOMATIC WRITING
RX121 BANISH PAIN
RX114 BECOME A NEW PERSON
RX210 CHAKRA BALANCE
RX134 CHARISMA
RX126 CONCENTRATION POWER PLUS
RX115 CREATE WEALTH
RX135 DO MORE IN LESS TIME
RX106 DREAM SOLUTIONS
RX119 FEEL SECURE NOW
RX117 HEALING FORCE
RX211 HIGHER-SELF
RX125 HOW TO DECIDE WHAT YOU WANT
RX127 INCREDIBLE SELF-CONFIDENCE
RX128 INTENSIFY CREATIVE ABILITY

RX122 LOVE MYSELF
RX207 MENTAL TELEPATHY
RX201 PAST-LIFE REGRESSION
RX202 PAST-LIFE THERAPY
RX111 PERFECT WEIGHT, PERFECT BODY
RX132 POWER & SUCCESS
RX101 POWERFUL PERSON
RX116 RADIANT HEALTH
RX206 REMOTE VIEWING
RX118 RIGHT-BRAIN SOLUTIONS
RX110 SATISFACTION & HAPPINESS
RX112 SLEEP LIKE A BABY
RX123 SPEAK UP
RX133 SPEED READING
RX205 SPIRIT GUIDE SESSION/MASTER SESSION
RX107 SUCCESS & EXCELLENCE
RX120 SUCCESSFUL INDEPENDENT LIFESTYLE
RX109 TAKE CONTROL OF YOUR LIFE
RX130 TENNIS
RX104 THE GOOD LIFE
RX129 THE POWER OF PERSISTENCE
RX113 THE UPPER HAND
RX136 ULTIMATE RELAXATION
RX124 WEIGHT LOSS
RX208 WHITE GOD-LIGHT MEDITATION
RX108 YOUR LAST CIGARETTE

CREATIVE VISUALIZATION TAPES

V107 ASCENT OF THE EAGLE
V103 BLOSSOMING ROSE
V101 EMERGENCE INTO THE LIGHT
V102 ISLAND SUNRISE
V105 RECONSTRUCTING THE CABIN
V111 TEMPLE OF SILENCE
V108 THE ALCHEMIST
V104 THE DANCING FLAME
V112 THE GODDESS
V106 THE LIGHTHOUSE

V109 THE MAGIC MIRROR
V110 THE MANSION OF BEAUTY

SOUNDS OF NATURE MEDITATION TAPES

M303 CREEK IN THE FOREST
M301 FOREST SOUNDS
M302 GENTLE RAIN
M300 OCEAN WAVES
M304 REFLECTIONS OF REFLECTIONS

WORDS OF WISDOM LECTURE TAPES

L802 BATTLE FOR YOUR MIND
EX103 CHANNELING PSYCHIC INFORMATION
L801 CRITICAL THIRTEEN
EX102 END TIMES & ARMAGEDDON
L110 ENTITY ATTACHMENT
L111 FIFTY PRIMARY UNIVERSAL LAWS
EX104 HOW REINCARNATION REALLY WORKS
L805 HOW TO BREAK THRU SUCCESS BLOCKS
L804 HOW TO END SUFFERING/ATTAIN PEACE
L806 HOW TO FIND A LOVING RELATIONSHIP
L108 PAST LIVES, PRESENT PROBLEMS
EX101 PREDESTINATION
L803 STOP SELF-DESTRUCTING
EX105 WALK-INS/CALIFORNIA FALLING
INTO THE SEA

25 BEST WAYS TO ...
TAPES WITH SUBLIMINALS

E102 BECOME A $UCCESS
E108 BOOST YOUR BRAINPOWER
E116 COPE WITH DIFFICULT PEOPLE
E111 CREATE HAPPINESS
E112 FIND LOVE/A SUCCESSFUL RELATIONSHIP
E107 IMPROVE YOUR LOVE LIFE

E105 INCREASE ENERGY
E103 INCREASE SELF-ESTEEM
E101 LOSE WEIGHT
E117 NEGOTIATE WHAT YOU WANT
E109 REDUCE CHOLESTEROL/HEALTHIER HEART
E106 REDUCE STRESS
E104 REVERSE AGING
E113 STOP SMOKING
E114 SUCCEED IN THE '90s
E110 THINK YOURSELF HEALTHY

CHILDREN'S PROGRAMMING TAPES WITH SUBLIMINALS

C004 CREATING GOOD STUDY HABITS
C003 DEVELOPING ATTENTION FOCUSING
C005 DEVELOPING CREATIVE ABILITIES
C006 IMPROVING READING ABILITIES
C002 OVERCOMING BEHAVIORAL PROBLEMS
C001 OVERCOMING LEARNING DISABILITIES
C007 STRENGTHENING YOUR IMMUNE SYSTEM

COMPACT DISCS

CD101 ATLANTIS: CRYSTAL CHAMBER
CD102 MANY LIVES AGO
CD103 SHARED BLESSINGS

NEW AGE MUSIC

AM122 AEROBIC EXERCISE MUSIC
AM144 ANGELS OF THE SUN
AM139 ASCENSION TO ALL THAT IS
AM140 ASTRAL MASSAGE
AM138 ATLANTIS: CRYSTAL CHAMBER
AM145 CELESTIAL BRIDGE
AM125 CHANNEL FOR THE LIGHT
AM109 CRYSTAL CAVE

AM132 CRYSTAL GARDEN
AM114 DAWNING OF THE NEW AGE
AM110 DREAMS OF ATLANTIS
AM116 DREAMSCAPES
AM124 EAST OF WEST
AM113 ENTRANCE TO THE SECRET LAGOON
AM117 GALACTIC ODYSSEY
AM118 HIGHER-SELF RENDEZVOUS
AM123 IN THE CAVERNS OF YOUR MIND
AM115 JOURNEY OUT OF BODY
AM119 JOURNEY TO EDGE OF THE UNIVERSE
AM135 KEY WEST AFTERNOON
AM106 MANIFESTATION
AM129 MANIFESTATION OF THE PYRAMIDS
AM141 MANY LIVES AGO
AM120 MYSTIC MEMORIES
AM137 NEW AGE GREGORIAN CHANTS
AM126 NEW AGE MUSIC SAMPLER I
AM133 NEW AGE MUSIC SAMPLER II
AM131 SEARCH FOR UTOPIA
AM128 SEDONA SUNRISE
AM134 SEDONA VIDEO SOUNDTRACK
AM143 SHARED BLESSINGS
AM111 SKYBIRDS
AM112 TEMPLE IN THE FOREST
AM136 THE ETERNAL OM
AM127 TRANSCENDENCE
AM130 TRANSITION
AM105 UPPER ASTRAL SUITE
AM142 WHEN I BECOME THE WIND

VIDEOS

VHS111 ACCELERATED LEARNING
VHS118 ACCOMPLISH YOUR GOALS
VHS132 ASTRAL PROJECTION & REMOTE VIEWING
VHS110 ATTRACTING LOVE
VHS131 AUTOMATIC WRITING
VHS109 CHAKRA BALANCE
VHS122 CHANNEL FOR THE LIGHT

BOOKS

Dick Sutphen is a masterful communicator who has established distinguished careers in advertising, brain/mind technology, publishing, and seminar training, in addition to writing several best-selling books. From 1958 to 1975, he received nearly 200 awards for the outstanding advertising he created for major ad agencies and for his own creative service organization. His clients ranged from industry giants, such as 3M Co., Betty Crocker and Texaco to political candidates and cities.

Effective advertising is the power of persuasion. In 1971, Sutphen sought to expand his understanding of this power by studying brain/mind technology and hypnosis. This interest became a full-time career in 1976 when he began conducting human-potential seminars and established **Valley of the Sun Publishing** to create and market the first prerecorded hypnosis tapes. Today, such tapes have become a booming industry, of which Sutphen's company is a leader, offering over 300 titles.

Over 85,000 people have attended seminars personally conducted by Dick Sutphen. His **Professional Hypnotist Training** is heavily attended by medical professionals seeking to incorporate Dick's unique techniques into their practices.

Having written many professional advertising and self-help books, Sutphen's publishers include Simon & Schuster Pocket Books, McGraw-Hill, W. H. Allen (England), and Valley of the Sun. Dick is often a featured speaker at such conventions as The World Congress of Professional Hypnotists and The California Council of Hypnotherapy. He has appeared on over 500 radio and TV shows, including **Phil Donahue, David Susskind, Good Morning America,** and **NBC Tomorrow.** Sutphen lives with his wife and children in Malibu, California.

FREE SUBSCRIPTION

Write for a **FREE** subscription to our exciting publication, *Master of Life*. It contains news and exciting articles on human-potential, self-change and self-exploration subjects, in addition to complete information on all Sutphen Seminars and our line of over 300 audio and video tapes: hypnosis, meditation, sleep programming, subliminals, and one of the country's largest lines of independently produced New Age music.

Valley of the Sun Publishing
Box 38, Malibu, California 90265
818/889-1575

Master of Life Manual

Past lives to human-potential principles—a complete manual of the basic communications to begin to create your own reality ... NOW! Sample of Contents: The 11 Basic Human Rights—demand them for yourself and give them to others. + The Universal Law of Resistance—how it works. + How beliefs destroy experience. + All problems between human beings are rooted in fear, but did you know there is ONLY ONE FEAR? + Conscious detachment—how to make any relationship work. + Karma and conscious/subconscious misalignment. + An underground best-seller, more than 145,000 copies are in print. Paperback, 128 pages—$3.95.

Enlightenment Transcripts

Dick Sutphen has packed a tremendous amount of awareness into this book, which takes up where **Master of Life Manual** leaves off. Thirty-six fast-reading dialogues simplify complicated concepts so they become more easily understood. Sample of Contents: Science has proven that energy can't die—and you are energy. + The concept of expressing unconditional love made workable and logical. + The four questions to ask yourself in order to let go of fear-based emotions. + Sex and spirituality. + The three kinds of guilt and how to resolve them. + That which you resist, you become; a Universal Law people prefer to ignore. Paperback, 128 pages—$3.95.

Sedona: Psychic Energy Vortexes

Sedona, Arizona, is the location of a vortex energy center that enhances all psychic abilities. Thousands of people have had incredible metaphysical experiences here, ranging from direct contact with spirits, visions, and healings to strange manifestations and effects.

There are four power spots in the world; Sedona is the primary power center and lies on ley lines connecting to Stonehenge and other spiritual power places in the world. A fascinating exploration of psychic abilities and the unlimited power of the mind, the book contains maps, directions, important information and warnings, and the stories of some of the visitors. 180 pages, trade-sized paperback—$7.95.

Lighting The Light Within

A collection of the best of Dick Sutphen's writings, gathered from his books, tapes, appearance talks, seminar lectures, and magazine articles, offering a unique, condensed understanding of metaphysics. He takes the "cosmic foo-foo" out of metaphysics and shows how logical spiritual principles network perfectly with science. His goal is always to assist you to let go of the fears blocking you from attaining your full potential to express unconditional love. Sample of Contents: The 20 primary universal laws. + Your earthly purpose. + Predestined directions. + How to end suffering and attain peace of mind. + Much more. Paperback, 128 pages—$3.95.

The Star Rover

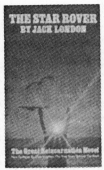

Jack London's classic novel about reincarnation — the fictionalized story of the out-of-body and past-life experiences of turn-of-the-century California outlaw Ed Morrell. Sentenced to life imprisonment, he was subjected to years of cruelty and torture. Through self-hypnosis and astral projection, he survived and was released as the result of a documented series of psychic occurrences. Author Ruth Montgomery's guides view Dick as a "walk-in" of Morrell. In **Past Lives, Future Loves**, Dick explores the Sutphen/Morrell connection. Includes a 23-page epilogue by Sutphen that provides previously unpublished information. A quality, trade-sized paperback, 348 pages —$8.95.

Past-Life Therapy In Action

Past-life therapy is becoming more commonly accepted, not only by the general population but by the brain/mind professionals as well. No one can deny the life-changing results attainable with these techniques. When your past-life subconscious programming is out of alignment with your present-day conscious desires, you will have conflicts, hang-ups, phobias, or undesirable physical/mental effects

in your life. In this book, Sutphen and Taylor explore many exciting case histories, carefully showing the **cause,** the **effect** and the **karmic lesson**, providing human-potential awareness of how to rise above the undesirable effects. Trade-sized paperback, 144 pages—$7.95.